THE GOLDEN YEARS

OF THE FOURTH DIMENSION

. . . .

Western Literature Series

★ ★

★

BOOKS BY KATHARINE COLES

The One Right Touch: Poems (1992)

The Measurable World: Novel (1995)

The History of the Garden: Poems (1997)

The Golden Years of the Fourth Dimension:
    Poems (2001)

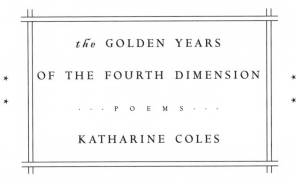

the GOLDEN YEARS

OF THE FOURTH DIMENSION

· · · P O E M S · · ·

KATHARINE COLES

UNIVERSITY OF NEVADA PRESS : RENO & LAS VEGAS

Western Literature Series

University of Nevada Press, Reno, Nevada 89557 USA

Copyright © 1995, 1996, 1997, 1998, 2000, 2001

by Katharine Coles

Manufactured in the United States of America

Design by Carrie House

Library of Congress Cataloging-in-Publication Data

Coles, Katharine.

The golden years of the fourth dimension : poems /

Katharine Coles.

p. cm. — (Western literature series)

ISBN 0-87417-480-5 (alk. paper)

I. Title. II. Series.

PS3553.047455 G6 2001

811'.54—dc21   2001001516

The paper used in this book meets the requirements of

American National Standard for Information Sciences—

Permanence of Paper for Printed Materials, ANSI z39.48-1984.

Binding materials were selected for strength and durability.

FIRST PRINTING

10 09 08 07 06 05 04 03 02 01

5 4 3 2 1

FOR CHRIS,
and the golden years

CONTENTS

ACKNOWLEDGMENTS

The author gratefully acknowledges the following publications, in which some of these poems first appeared:
"Eve," *Antioch Review* (Fall 1997): 458; "Shock," *Ascent* (Fall 1996): 46–48; "Advent," *Ascent* (Spring 1997): 41–43; "Abelard to Eloisa," *City Art* 4, no. 1 (Spring 1999): 28; "Immanence," *Seattle Review* (Fall 1998): 14; "Rocca Maggiore," *Paris Review* (Winter 2000); "The Moth," *Poetry* (February 1997): 260; "Wishing for Winter," *Prairie Schooner* (Fall 2000); "The Buffalo on Antelope Island," "Heidelberg Castle, from the Philosophers' Walk," *Red Rock Review* (Summer 1998): 99–104; "Judgment," "Petit Opéra," "Specific Gravity," *Red Rock Review* 1, no. 7 (Winter 2000): 103–112; "Stereopticon," *Quarterly West* (Autumn/Winter 1995/1996): 180–182; "Galileo's Finger," *Quarterly West* (Autumn/Winter 1996–1997): 124–126; "The Need for Science," *Weber Studies* (Winter 1997): 137–140; "The Golden Years of the Fourth Dimension," *The Sacred Place*, University of Utah Press, 1996.

Many of these poems arose from a long-term collaboration with Maureen O'Hara Ure, whose support, keen mind, friendship, and inspiration I gratefully acknowledge as a powerful force behind this work. "Cat's Cradle," "Galileo's Finger," "The Buffalo on Antelope Island," "Specific Gravity," "Sublimate," "Immanence," "The Moth," "Eve," "Abelard to Eloisa," "Heidelberg Castle, from the Philosophers' Walk," and "Rocca Maggiore" appeared in "Natural Histories," an installation that resulted from this collaboration. The installation was made possible by help from the Utah Arts Council, the Salt Lake City Arts Council, the University of Utah Undergraduate Research Committee, and the Salt Lake Art Center—

especially its director, Ric Collier. Another powerful force behind the creation of these poems is my editor, Trudy McMurrin, who always has my love and thanks.

I cannot overstate my debt to Michio Kaku and his wonderful book, *Hyperspace*, from which this collection's title and most of its epigraphs are taken. *The Golden Years of the Fourth Dimension* was composed largely while I was reading Kaku's book, and many of these poems respond directly or indirectly to the ideas so elegantly presented there.

THE GOLDEN YEARS
OF THE FOURTH DIMENSION

. . . .

Part One

★ ★

★

STEREOPTICON

If superstring theory is right . . . the 10-dimensional infant universe may have
split into two parts an instant after it came into being: one part became imbedded
in the familiar three dimensions of space and one of time, and the other part was
retracted from view, so that the six remaining dimensions became hidden in
string-like entities of almost infinitesimal size.
— MALCOLM W. BROWNE, New York Times Book Review

1. Mozart on the stereo.
                    Diminuendo.
                              In my rear-view mirror
the mountains kept shrinking, until they vanished

behind the green Midwest. On my way
from one edge to another. The map's unruffled line

translated into that range, rolling hills,
a steeple on the horizon becoming

a white-faced church, a blue door. A painting
you could walk right into, though I remember driving,

my windows up, through scenery
under glass
              not warped,

like the instrument's lens, which lets me squint
into, from, a projected place and time: two perspectives

bent into a single three dimensions.
                              Another version:
I did stop at that church. I stood,

hair blowing into my eyes, patient
for the click. Wind rustled the corn like fire,

swept the fields and moved on, eddied
in ravines, caught remote islands of trees.

2. If you looked in my eyes, you'd see it
curving there, even now, where mountains ring me
with slow violence. Another backdrop.

If a voice could open a pane of air
as a hand swings a window wide, moves the lace
curtain—even under pressure, the world yields

more of the same: a melting view, black
trunks of trees turning around, turning
into a house, paper, words, each a knot

I can set my foot on, to rise through
branches, the green roof lifting leaves
into an ether slick and hard with light.

3. Which also yields. The mind keeps moving
out, dazzled by space. Sun on snow
sends knife-flashes into my eyes
years later.
                    The tip of my tongue,

my mouth's arch of bone, echo spaces
I haven't yet imagined; a universe, withheld,
melts chocolate into breath. Cold morning.
The crossing guard takes me, at this distance,

for her own, and lifts her orange flag to save me:
a car brakes on ice, slides sideways into a drift
of speed. The future;
                    years burning;
                                    the shallowest time
accelerates so fast it leaves me breathless

4. here: white walls, corners, candles
half burned down, breakfast dishes left
on the table in the other room. A cloth,
deepest blue, sends us into evening,
where we bow our heads to what is

beyond us: the stereopticon, through which
these shapes leap onto the wall, present
a kind of depth. A train arrives. My great grand-aunt
reaches her hand through half a century.
I want to touch it. Her other hand smooths

overskirts of lace, through which light
casts its ornaments. The distance
between us. I have her name, resolve, something
of her eyes.
            I lay her down, dissolved
again to double surface, grains of paper, and reach

5. for the scissors and tough-stemmed grapes.
Twilight presses the window, but leaves
twine the sill, send shadows, tendrils
through the screens. A thousand tiny lives
tick in those spaces, too small
to see. Water poured too quickly trembles,

holds the brim. I want these things
to charm what already opens my throat:
vastness giving to vastness, fire, and dust.
Ignore the blade, how within its surface
our faces flicker; flesh keeps
holding perfect tension,
                              until it breaks.

CAT'S CRADLE
June 6, 1994

Waking from its dream, the mind leaves
an empty beach, a grey storm of surf,
an empty body rolling in the waves.
A white hand arcs and dives, the sea unnerved.

An empty beach, a grey storm of surf:
flesh and motion, too far gone to save.
A white hand arcs and dives, the sea unnerved.
In simple present, *now*, our hands weave

flesh and motion too far gone to save.
A loop of string, a shape suggesting form,
a simple present. Now, our hands weave
an airy cradle, woven and reborn

in loops of string. A shape suggests a form
that's vacant, offers emptiness to love,
an airy cradle rewoven and reborn.
Imagine a universe that breaks in waves—

it's vacant, offers emptiness, or love;
here, invisible within this space

imagine a universe, its broken waves
withdrawn, so we can't touch its other face.

Here, invisible within a space
I pace as if I know it, what unfolds,
withdrawn beyond my touch? Another face,
matter withdrawn from what matter holds.

I pace as if I know what folds
behind the world that presses us: green leaves,
matter withdrawn from what matter holds
against us — dusty sunlight, a stiff breeze.

Behind the world that presses us, green leaves
its glimmers, simple actions of the mind.
Against us, dust and sunlight, the stiff breeze
may be God, may be only blind

glimmers, actions of a simple mind.
Waking from its dream, the mind leaves
what may be God, what may be just a blind,
empty body rolling in the waves.

THE NEED FOR SCIENCE

for Chris, on the anniversary of moving into our house, August 14, 1989–1994

1. Invisible Weight

[I]f appearance and essence were the same thing, there would be no need for science. —MICHIO KAKU

Or microscopes, telescopes, steam machines
for stripping wallpaper—that bathroom,
navy blooming with pink

irises the size of my head—poetry, news
analysts, physicians, the FBI, dating
services. The perfect match, we meant

ourselves for one another, at first sight
(allowing for the collapse

of what seems *no time*), so made
ourselves over,
                took

each other's measure, lip-to-lip,
did not count seconds speeding up

our heartbeats, washing
over our bodies—the past emptying

out the future's rush and roar
dimmed by the sound of our breathing, the hum
of his old air conditioner, heaved

down one set of stairs, up another. Every touch
left its smudge, its slow, cumulative,

invisible weight.

We'd had to wait

an age for each other. And we had
what still looked like forever.

2. Visible Weight

*By simple rotation, we can interchange any of the three spatial dimensions.
Now, if time is the fourth dimension, then it is possible to make "rotations" that
convert space into time and vice-versa.* —MICHIO KAKU

If I could turn a Kenmore washer into time,
I could rotate it through this door
elaborated by a Victorian mind

that wouldn't have conceived
such machinery. Or
that I would want it, a hundred-

some years down the line. I have
misread again, willfully,

not only science, but history—
it is so hot, and the washer

so unwieldy in its space,
who could blame me for reducing theory
to mere machine?

The physicists,

clucking collective tongues,
precisely measured. Their voices

take just so much space in my mind.
Call it *x*. In time,
they'll shrink to nothing, small matter

converted into energy I could use, now,
resting my back against dusty woodwork,

while my physicist watches over his glasses.
All before we married. He considers

matters of space and time,
                              machine
versus merely human mind. Counts
complications. The move, the wedding: all

sooner undertaken, sooner finished.
Since then, we've learned a thing or two,
have buried friends we held,

a mother who held us. We recover
nothing: holding each other, we hold

each other's absence. We are turning
into the past. In retrospect,
I would prefer to take my time.

3. Anniversary

Newton, writing 300 years ago, thought that time beat at the same rate everywhere in the universe. . . . However, according to special relativity, time can beat at different rates, depending on how fast one is moving. —MICHIO KAKU

Another orbit finished. Recollections
past, or passing, by the time we mark

a heartbeat, a line—*anniversary*
and *universe* both contain that turn,
that walking rhythm. Long

days rush us through
the universe, the universe
through us: another year, the nightly throb, his pulse

against my pulse, starlight's insouciant wave
rippling the screen. The blinds

flap in arid wind, the heat wave
we confuse with
five years back, summer

beating down two years before,

                            repeating
a house-of-mirror's trick. Hell,
it's only time. The day we fell

it must have seemed to him I stood still,
one hand resting on a book, composing

my response; but my mind moved
so fast he'd have seen its blueshift
if it were a star, he a stargazer watching

space collapse between us. *It must
have seemed to him*, but I don't know.
We move through different spaces, different times,

the same space and time differently—
I love the distances, roughnesses,

rotations, odd warps and woofs
we travel to touch each other.
                              On my birthday
two years after we met we moved in here;

in between, a love at first sight
took two full years to ripen,

then was there. It is my birthday today.
How long has it been? we ask each other. *Yesterday,*
forever. The bathroom's eggshell walls

needing paint again, a couch gone dingy, paired
chairs we sit on, staring
into space: all collapse, spin

into mystery. I still love,
over time, even the damage
time has done to him, though, minute-

by-murderous-minute, he looks the same;
though we move so fast
we only seem to have stood still.

GALILEO'S FINGER

Palazzo Castellani, Museo di Storia della Scienza, Florence

1. Missing what we're pointed to, we find
that flooded crypt, water

so clear we have to pay to see it
ripple its surface over our tossed coins. Wishes

we lean into, faces glinting. As toward
that village museum's great painting—a Madonna,
looking, in spite of all the stars, plain

worn out. Outside the museum, gypsies
picking another tourist's pocket curse us,
and probably our children, inventing a future in words

we can't understand, even then.
                                    One more glass of wine
at lunch. We can afford

to view art and artifact, mazed streets, their universe
at our feet. Wonder, Why his finger? All bone,
pointing up through its gilded egg

to receive the second spark, or to raise
eyes to our moon's irregular face—
though even he couldn't believe the moon

might drag an ocean, so much
loose skin, across the bone—or to that more distant moon
his eye rode the lenses to fix in heaven. The finger

could be anybody's, dead
three hundred years. The nail, grown long

after the heart stopped, nudged
simple movement into revelation: piano inclinato,
pendulum, thermometer. Conceived in numbers,

that older world revealed itself
to steady eye and hand—even telescoping

the long view, he couldn't see the past
our flying instruments would measure
in his name. A machine angled just

right for once, its parachute blowing
open, carrying us so quickly, burning
our shapes on the dark.
                              Einstein's brain, too,

is under glass, sliced, scalloped, mapped, continents
thin as lenses blurring the world, clarity

we long for. An idea he had
to close his eyes to see. In the next room, planispheres,

astrolabes, armillary spheres elaborate still
unknown continents, the earth
around which planets swing, stars fixed

as if a thing of beauty could make knowledge
visible. Or the other way around. Gorgeous
wrong turns rippling into life.

2. Would you renounce
the universe to live?

                              I'd never thought
imagination made time. Harpsichords

devising numbers into sound. The player piano's
works, which don't, for us, but in

the ear's private drumbeat. Jeweled
faces made austere by silence. We can't help

nostalgia. Needed to know the microsecond
the second we could, though my cheap watch,
set daily, registers my mobile touch. Time,

more than ever, is beyond me. Perhaps those gypsies
should just spend what they have. Stop desiring.
Stand before these scales, where *just* means just

precise, relative to
some other human weight. Would you renounce
truth for grace? Different weights

falling at the same rate. Through time, the gypsy waved
her newspaper; her daughter's fingers worked

the tourist's pockets, his zippered pouch, so light
he never felt them. She had the touch. The timeless

face of a Madonna. And once
she had his wallet, needed it more

than when it was mere idea. She held
so tight. *And yet it does move,*
Galileo said, having just surrendered

all he knew. The gypsy said, *It's mine.* Determined:
we, products of our time, would help him

wrest it back. By now, from his porch in Tennessee,
the tourist has recited the event, changed it

with each measure, justified. And she,
making time, ticked her teeth, plunged into
the future's ancient atmosphere, skirts blown adrift.

The Galileo Probe, December 1995–January 1996

THE GOLDEN YEARS OF THE FOURTH DIMENSION

for Elaine Smith, November 4, 1934–July 14, 1994, and for Michio Kaku

For uncounted centuries, clergymen had skillfully dodged such perennial questions as, Where are heaven and hell? and Where do angels live? Now they found a convenient resting place for those heavenly bodies.
—MICHIO KAKU, *Hyperspace*

1. I confess, I am like them—fabulists, theosophists,
readers of the moon, of tabloid science, *fin de siècle*
purveyors of raps and voices making matter

of the thinnest air. Oh, the drapes and flashing lights,
the glittering fissures.
                        Elaine would have loved it—to imagine
they *are* tiny, those angels, enough

to make a world of a pinhead, a ballroom
vast enough for the cosmic dance, the space

they inhabit smaller than an atom.
A simple one, say hydrogen. The lightest thing
we're made of, or so we thought. Though now

I imagine angels crawling my blood, heavenly,
metastatic, working from the inside
out. The difference

they can make to us, those dimensions, higher
but so minuscule we contain them. Say
the atom splits. Tears

open into our space: a rent
so bright we can't look into it, can see
only fallout, aftertraces of dust

decaying into poison. Not the moment
angels glance out at us, surprised

at the rupture, as if
by a camera's candid flash projected

into cubist life. Still
invisible, to the human gaze. Dropping nail files.
Rubbing sleepy eyes. You don't believe

they spend their time singing, thinking of God?
We couldn't know, then, what we'd unleashed.

2. But this was long ago. And before that, the century's
extravagant turn, its desperate theologies—

what to save in the face of Darwin? Of Einstein?
Of Lenin?—who said the fourth dimension

was no place for revolution. Oh, yeah?
my dead friend may be saying, sallying forth

to see what needs her fixing. Others
proved him wrong—and they
kept coming.
                    So, matter mutates

into energy after all—not only radium, fixed
in Madame Curie's gaze. But our simpler flesh,
compressing under time. Today,

I listen to eulogies full of sweetness, weighted
with what seems unspeakable—which is relative,
after all—and remains

unsaid. Elaine's image decays
before her body, weakened inside-out. Life

into half-life. In the end, we take
what we can get: love, which serves

itself. Mine, too. I loved her
wired tight and angry. "Shit,"
she'd say, "tell it like it is." A quality

her sister-in-law, smiling, pronounces
to get it off her mind. Elaine left her

helpless.
        Elaine
scrubbed my stove, or leaned against it,

waving the sponge at me. Scrubbed
my mother's kitchen floor, and later mine,

on her hands and knees, backing up,
talking, into afternoon. Loved

what never earned her love,
and she knew it. Ran, one day,

over her lawn to greet us, right into traffic,
waving her hands. Every car swerved,

missed. She had a month to live.
                        That physicist,
leaning, a century ago, over radium

decaying, breathed its dust, inhaled
the last century's air, expired
it into this one, where

the Czar is overthrown
in solid three dimensions, where we mourn
Elaine in a suburban chapel. I touched her hand

one last time, though it was cold and still,
at least to my inadequate eyes,
and the mortician had sewed her mouth

into its dourest frown. Someone said
how beautiful she looked, and at peace.

3.   I never saw
the Czar, except in photos, a figure
made of light.
                    There is no difference

between time and space, between
loving her from week to week
as she was, sponge in hand, and loving her

now, maybe across unspeakable dimensions,
translated into spirit by desire. Except

to the human heart. It's all
the spiritualists wanted, somewhere to go
when they're finished here. If she's anywhere,

she's energy, not matter. Beyond
our reach.

Even that moon
for so long was so distant

it might as well have been
a *heavenly* body. As of this month

twenty-five years ago, men have walked
its surface, have seen the earth rising

into a moony, blue and white romance.
Twenty-five years is nothing, except to us.
Only six more: another millennium, a new spin

on events she wouldn't have wanted to miss.
So many scientists want to tell us

this is no mere metaphor. But
even the scientists talk of light
as a lucid ocean beating our shores;

even they, knowing *so much*, know
only the approximate, the magic change.
Light may be

vibration from another dimension.

Or not.

The scientists,
too, are fabulists, in love
with whimsy and invention. In such a universe,

numbers are pure only to the mind.
And the rest of us, who have believed

in what is beyond us, have always thought
it must be, after all,
too large, not too small, for us to witness.

THE BUFFALO ON ANTELOPE ISLAND

The Great Salt Lake, 1994

There are no buffalo on Buffalo Point—
just the stunning elevation
into midsummer heat, rock, and scrub.
And the signs, Bison Can Be Dangerous:
though the buffalo are penned,
dusty and bored, to save us from our desire

to approach their wildness, and touch.
A horn could tear a body throat to crotch,
has gutted tourists before, who buddied up
for the perfect snapshot of themselves
and their closest friends—the buffalo.
Bison can be dangerous, can overtake you

before you know it, tiny hooves moving
a ton of flesh into unstoppable force
that carried them even here. Not these,
trucked across a man-built causeway,
but their authentic ancestors. They appeared

from receding, prehistoric waters
by transmutation. No, they swam,
floating all that bulk upon the brine.
Such a blue buoying the barren hills,

no wonder those first explorers thought
they'd reached the ocean at last, overlooking salt-
dazzled water, horizon of island and mist.
Seven hundred miles still to go.
There are no antelope on Antelope Island

anymore, as far as the eye can see.
The buffalo inhabit what we give them:
one word, another, a bare island, their pen.
Bodies at rest. And the buffalo
on Antelope Island are boring,

as only the truly bored can be. They lack
our imagination. Even we
have trouble investing them with inner lives,
the outer is so meager, so dusted over.
We hang from the fence into the cloud of flies

where the park's contained them. Next to us,
a boy pelts them with stones. If they could rise
and eye us from their own unfenced
place and time, that would be
another story. Wouldn't it? And we,
believing in the signs, would keep our distance.

JUDGMENT

After Jan Van Eyck

I will also send the teeth of beasts upon them, with the passion of the serpents of the dust.

1. *The Crucifixion* and *The Last Judgment*

So I've stopped believing
I'll live forever, in spite
of rising's afterthought, this

too-willing sacrifice projected
in brilliant, tiny strokes. A symmetry,

no more. I'm living
happy-ever-after: the painter had—

and I, so modern—*matter*: his, the soul's;
mine, the body's; and the demons
always have too much

fun, spiny, delectable fingers
dismantling torsos, splitting

(I hadn't gone so far
as to picture *limb-from-limb*)
the pelvis open, a halved peach, which

has always burned, along
the natural breach. Right to the navel. Rendered
distinct, each

crimson sinew tearing, each mouth
opened into the howl. No matter

I can't hear it—
            Heaven's also
out of reach, angels and citizens ranked

shoulder-to-shoulder against the maw
lapping at their toes, foreheads blaring

reflected light. Not they
who draw the eye, whom the brush strokes

into fire whose wind pulls me
closer. Wanting to look down, half
hoping, always, for the fatal slip.

2. Interlude: The Musical Instrument Room

Downstairs, the medieval chessboard's
king of wood and gilt and bone. Which,

to end, the loser knocks
over, to keep it

from the taking. Or that pendant—lovers
naked under Death's gold-sparked eyes.

Death itself, languishing in the form
of one saint or another, of Christ

draping a marble box the body's size, hand
curled, palm-up, around the air. Still

preserved in marble, immutability a final
relaxation.
                    Gilded harps, each
curved to sail to heaven, delay

my descent into Sunday crowds, the A-train's
weight of summer wind. Currents

ripple down one gallery, up another,
through horns in the shapes of fishes, dragons,

jaguars teething the air; inlaid
lutes tuned to summon gods;

this gyo, cat-shaped, its spine raising
purrs to end a Confucian hymn—each note,

under glass, trembling
toward us through the gallery's hush.

3. The Missing Panel

In a few years, the Renaissance. Lost:
long hours in the middle. Birth.

Death. Rebirth. Seconds and days
finite because no one attends
their fracturing. For now
Van Eyck turns me

on suffering's wheel, though joy
and heartbreak accumulate in waves,

wash through me, unexpected,
then recede. Trivial, when demons

press their faces to
the world's one-way mirror.
                              The panel's absence
illuminates an altar: no caterwauling, boiling bones,

but the milking, the pulling
of beer from the tap. A woman

wears blue by a window overflowing
with sky she looks

beyond, to the vanishing
point. Her demons she keeps
private. A fire in the grate. She sees

another woman, cloakless, gathering
tinder left on the ground

so late. She sets down fancy
stitching to look out, past the frame.

4. Laying On

Only in Paris and Rome have I seen
such impractical shoes, so many women

smoking in the streets, so much
underwear showing. I am
from the provinces, where

dowdiness is no more dowdy, only
more naïve. Underground,
in the roar and crowd, children

too old to be sucking thumbs suck
away, asleep, piled on the bench; young girls

wearing beauty's armor—nothing else
could hide them—step through
stinking puddles; this man,

who only walks as if he's old, drags
a handled sack from Macy's, his stained coat

the house he paces. At the edge,
I believe myself
beyond him, until his eyes lift, take me in

a moment through which his hands descend
to press my head between his palms, his breath
a wind so tightly woven

I can't unravel it, so hold my breath
against it. His eyes keep mine

open when our lips touch. The train
rattles into the station, slows, sighs
to stop. We separate, take

separate breaths. No eyes
on us. All voices held, until
the door slides open on the rush and din.

SPECIFIC GRAVITY

A piece of sky
flutters loose—a jay

resolves himself before
a yellow flight of leaves,

settles into them, absorbed.
His blood denser than water,

he lifts into motion
lighter even than air,

into which leaves turn
to light, then to wind—

how this longing,
so precise, accepts

one answer: blood,
muscle, bone knitting

a singular shape. My heart,
anchored three-quarters up

in my body's water,
adrift, opens into

brilliant wind, and leaves.

CAUTIONARY FIGURES

. . . .

Part Two

★ ★

★

IMMANENCE

She hoped to be wise and reasonable in time; but alas! alas!
she must confess to herself that she was not wise yet.
—JANE AUSTEN, *Persuasion*

Probably any poem would have made her
feel it, any back-page political column or notice of arrest
or recipe for, say, eggplant roasted with garlic

or chocolate bread pudding or golden potato soup, or—
why not?—macaroni folded
silkily in its blanket of cheese, cauliflower nubbins

to lay on his tongue—all of which, reading
their spells, she imagines hovering over pots of,
spoon in hand, for him—nor

does it help to close the cookbook on her finger,
look out at snow blowing under a streetlight, winter
arriving to stay, flakes

now seeming to falter, suspended
starlike in their particular flights, now
driven across that bright eddy into the dark—so generalized

is her longing, nostalgia for a touch that, she must
confess, she has never yet felt,
except at the heart, however clearly

she remembers the way it entered her—
on an evening invented by summer's
deep, late light—for good.

SUBLIMATE

Fall's first frost. Moonlight
sparking the grass. Ravish of wind
behind which snow almost could
take its body from vapor, mere idea
in a corner of the sky,

                              under which,
beneath the window where my husband sleeps,
all that holds me is gravity, what
I will not think of, mathematical
promise in which I become

abstracted, mass and force, hinged
joints shifting under pressure, weather's
long release. All around
the leaves let go to fly,

                            and why shouldn't I
rise from dream into this cold, bare-
eyed gazing at the fleet
of stars? Time's wave curls beyond
what I can see, holds
its arc until it aches, and then

breaks into the fall—burning
gas, froth of atoms
fractured at their hearts—where

horizon returns to center. The past
looks just like the future
until that turn.
                          A dusting of soot.
                                        My heart
beneath my thin gown unfurls, closes
to recover. Flint and spark. I've breathed

his breath, let him harden
along muscle and sinew, taking my shape,
ice limning a window. In his arms tonight,
I kept my eyes open, held him

in mind as long as I could before
I closed them;
                          turning in my dream,
when I licked my lips I tasted
his salt left on my skin.

WISHING FOR WINTER

after a collage by Betye Saar

1. We see into the window, not through —
blizzard pushes its back against the glass.
A pair of dice. The glove he mislaid last
fall this time, bird's wing, a key whose
lock is long lost. Memory moves
the corner of an eye. These early snows —
did we really long for them? — blow
a note so true he finds it on the flute,
scales against it, as if the air could be
at once presented and withheld, frozen
under glass. Until the wind shifts
south. Fall again. Middle C's
a construct, not an absolute. Beholden,
we cup our ears, let go of perfect pitch.

2. We cup our ears, let go of perfect pitch
and take a wild run at high notes
we squeak onto, if we're lucky, or float
right by into wrongness, the voice's catch-
as-catch-can exuberance. My high G twitches
like a horse's tail. Not middle age —
I could never sing — but courage
brings us to extremity. What's out of reach
I'm willing to let go. And from the wreck
of wind-blown glass, life gone too soon,
this approximated song, gleams
a fragment, moment, one note plucked
and recomposed by light — wrong undone
and done again, more right than it seems.

3. Undone again, more right than it seems,
the year stretches out behind us, landscape
blurred by time, by wind-blown snow, or draped
with greenery, the winter's loss. A dream
builds itself all night beneath the scream
of wind, until singular voices merge
and we wake, shaking, with some urgent
message lost to sleep. We can't redeem
the past, though we relive it every night,
forgetting who is dead, who alive,
who ever had the right to ask such favors.
The mind, overgenerous, reignites
dead faces by its own imperative,
never minding who we've saved a place for.

4. Never mind who we saved the space for.
The house is full of ghosts, and better them
than emptiness, the long reach of autumn
echoing mere wind. I see his face, or
his dead mother's, though a turn undoes her,
turns her—all her instinct—back into him.
After all, it's just the consequence of time,
a conversation that holds us more
than we hold it, and holds us to the earth
not even long enough to form a word
before it takes us back, folds us inside.
No syllable on the tongue, no whole breath,
no partial breath can be heard,
no single letter permanently embodied.

5. No single letter. They're permanently embodied
in these obits, these notices and cards
saying *much loved, partner of, hard*

luck. No word was less inspired, though God
knows these were sincere, the form friends had
to follow at the time. Funny, to be
the ones who get all the flowers and pity,
to *stand in* this space left by the dead.
Filing by their coffins at the viewings
we stare as hard as we can, try to catch
the flutter of an eyelash or a sleeve,
some ghost of motion we must then refuse.
Because they've moved beyond us, out of reach,
they're no longer like us. So we believe.

6. They're no longer like us. So we believe
even against their absence in permanence,
each other's and our own, the long romance
stretching its golden future. Meantime, we live
dailiness—blue sky going grey, leaves
flying south, the north wind tuning up
to rail against the present. We fix supper,
setting out forks, then plates, then knives,
haunted by the day, by tomorrow,
by the coming winter—blowing hard—
that gives these days their edge, and their brightness.
Even our bones wrap a sweet marrow
some rough tongue could scour. We stand in the yard,
breathe the last warm wind. Just to notice.

7. We breathe the last warm wind, just to notice
weather bullying in behind, drafts
of notes to shatter glass. The snow will drift
against the porch, the empty wrought chaise
longues on the deck. We don't confess,
while we bring in pillows, fill the feeders

for the birds, how we like wind's beating,
how our blood rises, begins to race
against the turbulence, the tenor of storm.
That early winter was just a counterfeit,
false not in cold, in how hard snow blew,
but in endurance. Eyes, hearts, arms,
hold calm against the raging, contradict
what we see in the glass. We are not through.

THE MOTH

for Virginia Woolf

First, I am falling out of love,
falling giddy out of the gown I put on
to take it off, out of lace
and satin underthings, hooked and eyed

as if they were the object. I am
falling out of the lamplight
where those nothings lie, beautifully

empty puddles after rain—falling
out of that wave ticking constant
time against my heart. Out

of mind. There is
my last room, which I am
falling out of: only a lamp,

a clean desk, floor scrubbed bare,
one door to latch and lock, bed
sheets pulled tight enough to hold me
falling into sleep like a stone.

SHOCK

. . . *as the Physician cures him who hath taken down poyson, not by the middling temper of nourishment, but by the other extreme of antidote.*
— MILTON, Tetrachordon

1. So many ways
to die. Last night, I daubed

his arms with alcohol, prepared needles,
the serum slipping under the skin —
                                        how quickly
poison, taken weekly, grows so mundane

we had the TV on. A Million Square
Miles of Sunshine, the weatherman

announced. The west dried to tinder.
Later, leaving the hospital, we smelled

a premonition of smoke on the air.
For the moment, we were safe again,
never again in the same way. Trying

to build resistance. This morning,
we learned a neighbor died of it — flowers
and grass, pollen of trees: the body

protecting itself, so vehement
it killed: skin burning, allergy
shocking the throat closed

against what mustn't enter.
                                And what must.

A locked door away from antidote.
At that moment, or near to it,

the emergency room doors swung open
and my love walked out, on his own feet.

2. A minute.
                Fifteen.
                                No time.
Forever. I watch the news for rumor

proved: a telescope gazes back
ten billion years; a universe

erupts into time. Nothing
approximate, nothing like

dusty family albums: a minute
for the economy, five for weather,

for sports, fifteen to O.J.,
                        speculation
yielding to speculation—a wave of glass
fills the eye with stars, narrows sun

down to light dry leaves,
places a body

here, revealing traces, but cannot show
that explosion—
                  thirty seconds to fourteen firefighters

who could only lie down beneath flame.
Not quietly. Fire's

louder than any cry. Bones
crack into ash, darken the air

faster than we imagine. Say a lens
could carry you back, to see

yourself. Ten years ago.
Twenty.
        Half a century.

3.   Who'd guess
fire could move so quickly. That memory

could obliterate the past. That someone
would have made a silver tent
to lie beneath, so that fire burns

over you, if only it keeps moving.
It didn't.
        Today is clear

and cool, as if fall were coming on.
But late last night, the wind

carried smoke from fifty miles away.
A different fire. Only houses burned.
The wind blew ash of shutter, ash of sage —

evidence so tainted we couldn't read it—
hot and dusty, as if air might ignite.
Snapped the outside wires. They danced, alive,

sparking little blazes in the grass.
We are only hydrogen, after all.
We stood out on the street, holding hands,

composed, though ash fell onto our hair,
and sirens wailed toward us.

PETIT OPÉRA

for my mother

I believe that electric forces are definable only for empty space.
—ALBERT EINSTEIN *to* MILEVA MARIC

What's the point of falling in love nowadays anyway? It's such an old story.
—MILEVA MARIC *to* ALBERT EINSTEIN

1. The Secret of Empty Space

All

he got away with: what he could carry.
No television, no VCR. *Why didn't*

he take the stereo? I say, unhinged
by French doors' splintered daylight,

eyeing the mug he drank from at the sink,
still hearing courtship's cadence

in Albert's letters to Mileva before their marriage,
hers to him. My mother's house,

cool among old trees, is enough
to raise desire in anyone, in this heat,
who drinks at her kitchen sink

her water, gazing at her green yard,
wanting to smash something.
No way now

to warn Mileva a century ago, in love
with Albert, heavy with the work,

so we catalogue empty spaces:
a high-tech toaster mistaken
for valuable, up-

ended; camera, Walkman, two shelves
of operas. *Turandot*, my mother lists,
shutting her eyes to recover time

her throat opens around, and two —
no, three — versions of *Carmen*,
and that one, what was it called, about Dido;

two hundred disks, each heart-sized
surface coded with soaring, secret loves.

2. The Secret of the Stolen Songs

He entered space
into which his shape can't be translated,

becomes an appearing or vanishing blob,
a series of circles, lines
preserved on a fragment of glass. Even here,

in the West's dusty, devastating heat,
her disks shine like coin, the song of change
on the palm, the pocket's aria. He had

all the time in the world, no time; all day
the heat will not break, so tempers do
in houses and bars, under

the blare of sun, and now
into twilight, through which heat keeps

rising, radiant, from sidewalk
and street, brick walls guarding
respectable houses. Voices break,

heads break, kitchen crockery
shatters into the dark, and the police keep

not arriving, the only harm
to property, my mother's innocence. They'll file the case,
move on. Passing time, I skim

down the Yellow Pages, repeat numbers
for CDS, sold and bought. My mother

whispers operas at the receiver,
throat still humming. The next afternoon
at Discriminator Records—*He's there, what*

should we do?—Bill, behind the counter, opens
every disk, holds it to the lamp,

rotates it under his gaze. You'd never know
light unlocks music. *Those unheard*
are sweeter, he says, and laughs, though business

depends on the opposite. Still time. *This one*
looks okay. I memorize a face; Mother rifles

the stock; Bill turns another disk; the face,
beautiful, browses stacks over shoulders
not quite twice my width; I

look at my watch, watch the door, wonder
what I thought I could do. My mother

cruises the aisles, grey-haired
and harmless, if you don't know

better. Could we, among us, wrestle
him to the ground? A hundred-five and rising.

3. The Secret of Love

Albert wrote of science, Mileva of life.
Albert of love, Mileva of obstacles. In his letters,

he said, *our* work, *our* theories,
generous in love. Or ambition. What they carried. The gain

and loss of love: another story.
                          Later, we say, *no wonder*
he wasn't nervous. Robert, selling the disks,

has no idea, is green-eyed, tender,
doing a favor. Steve waits in the car

though I wouldn't leave a dog out in this heat.
Robert's secret is, he loves—or loved,

until an hour ago, or hoped sometime to love—
Steve, a three-week-old romance, whose name
will turn out to be *Guido*. I'm getting ahead:

his name, when the police
arm him out of the heat, cuff him, read

his rights, is Steve,
                              and can Robert
slip that name off like a glove, slip

new love off like a name, or is Guido
a different fellow? Who never

had a chance in this small town:
                              Bill, knowing
the tune of my mother's passion, sold her

all those songs; and, though
Steve points at Robert, Robert's best friends

with a friend of my mother's; Robert plays
the flute; and Steve—né Guido—wears

shorts too much more explicit
than his denials. Can nothing reorder
the heart's cadence? No underwear, no gloves—

Steve never counted
on my mother, who made the police, tired,
too busy for argument, dust each shard of glass

for prints. Even the police
are bemused by mother and daughter

getting their man. Another old story,
though not as old as love. What's in
a name? In Mileva's, now Einstein,

never on the work; in Albert's
growing digressions of attention, then of faith.
He didn't take the stereo because he took

the bus, music loaded into two
public television tote bags of my mother's.

He must have fit right in, but
for the shorts. Guido, we learn, gave

each new lover a month, then robbed and beat him,
then moved on. Not one,

Guido, to take his time, to drift
from passion to indifference. The detective says

they'll put him away for years. Robert phones my mother,
late at night, and weeps. That old song.

Even under leaves falling
into our hands, under the rough
skins of beasts we named together

and harvest now, eating
our words; under a sky

reeling overhead luminous
strange bodies to mark the season
toward our passing; even reading

heaven's signs traced
on our darkened hearts, I bear

my body as I never did
when I wore sunlight, only
skin I never considered, never

carried as loss. I did not know
the shapes desire takes—

so many!—on my surface, his eyes
on me all evening, pressing
a curve against my dress, fabric

I lean into, watching him
watching. His eye reforming

nipple to a berry for his mouth. Just that
hardening: flush
beginning deep, traveling sinew,

knotting muscle to skin—could I
bear the naked gaze?—so rich,

this gravity, longing
plucked from pure air. It thickens
under my hand, my belly where

his touch moves into me. I don't know how
I could have lived without it.

ABELARD TO ELOISA

After lines from Pope

I would take you from him, if I could.
I'm more man than any plaster saint —
how could that marriage be more true than ours,
though I gave you to it, sealed it with a ring?

More man still than any plaster saint,
I'll give you all I am. Dream the rest,
all already given. Though sealed by rings
I'll make you mistress to man again. Of love

I'll give you all I can and dream the rest.
I am not cold. If Eloisa loves
I'll make her mistress. Man again, I'll love
your image, steal between your God and you.

I can't be cold, if Eloisa loves.
In just such fires I have raised the altar
for your stolen image. God in you
is all the God I see, unholy joy

fed by fire it feeds. I raised the altar
you lie before to give yourself to Him,
all the God I see. Unholy joy's
made no more holy by the way it's spent.

You've lied before. You give yourself to Him,
to a marriage no more true than mine,
made no more holy by the way it's spent.
I would take you now, Child, if I could.

THE GOLDEN MEAN

For Anna Karenina

My baby thinks he's a train.
—ROSEANNE CASH

Anna, it's awkward, beginning
with apology, a confession
it's *not* funny, however
invented you are—broken
just where your author's

love broke the hinge
at the back of your legs, folding
you to beautiful ruin
at Vronsky's feet, his own,
so many pages before
you flung aside what little
they'd left between them: only

your body, antidote.
The rest given over
and over. Temptation *is*
everywhere—my love
could be next, his innocent
head bent to book, the image
of husbandly contentment
transporting me. Or

does he compose this
for me, to hide the plot
under his golden hair?
And I, better at reading
other hearts than my own,

might almost turn
betrayal's page, begin

living my confession,
get it behind me. Satan's
inkiest scales are limned—
so I hear—with gold.
And somewhere between
seductive void and glimmer,
black mood and night
so strewn with stars

I lift the book from his hands,
lead him out to the yard
to stand, head back,
eyes following the long
trail of milky light,
we walk this line we've learned

to call our lives, toward
a vanishing point—a distance
you've traveled, Anna,
in pages, measuring
plot's mathematical arc,
its dire cure. We call this
middle age, unable
to bear being past

such measures—though
you were. Would
we, if we could,
flip to the end, if only
to check the pages

against our progress? Fearing
any answer, extremes

we might be given to,
knowing this is not
our only possible life,
but, once it takes us, becomes
the one we get. The others
lost in transit. You
rode the rails into

perilous fate, and then
rode them out again,
conclusively. Never thought
to step back, to change
your ticket, change trains

in any one-window station.
Change even your mind,
if only they had left you
a mind to change, on which
to ride these pages out. It isn't
funny, I say again:
the power blip, a ghost,

wind from calm skies
bumping the CD player
on as I close the book, and you,
having turned from the world,

step from its platform into
the invisible, departing
with that train. My husband,

preferring your country's music—
Rachmaninoff, Tchaikovsky—
to any country song,
when Roseanne starts in, says,
What the hell is that?
and laughs. Are these really
golden years, or only

some middling way,
or worse, our dark age?—
so *living without*
contains its opposite,
as I live with

the one I've chosen, keep,
unlike you, choosing
(now, Roseanne turned
off, his head, still gilded,
bends to its communion),

but also out of almosts,
not-takens, ghosts,
abandoned chances, some
jostling my elbow yet,
plucking my sleeve, whispering
invitations—say

this one, whose creased lobe
my tongue recalls exactly;
or this one, whose hands
I might have let transport me,
might once have let lead me

to the rail, lead me
to say *I am lost.* I have
lost them all, having
given them up—though not
shaken them. I keep
hearing their voices, under
the chug and whistle of time,
this compartment made of lamplight
that carries us, swaying

to separate dreams.
Never so crowded.
Why Anna *again?*
he asks. He believes
you figmentary—you, Anna,
true as gold—as once
I believed us safe,
gazing through winter branches
at the progress of stars,

through the mullioned page
at the station sliding by,
the platform where you watched
train after train arrive,
board its passengers—
Anna, wave good-bye—
then pick up speed without you.

A king's foolish leap
into love—the castle burned
for better and for worse, for centuries.
Rubble now. Tourists, drifts of weather
before the colonnades. All for a girl
whose jowls the guides make fun of,
posing beneath her darkly painted eyes.

                                               Adam died

licking juice from his fingers. Eve
opening her hand. Millennia later
a woman starving for love in Guatemala
unrolls her sleeping bag in the plaza. Would you
for me? my love asks. Enough to give
a body to the burn we die for
every day of our lives. The rest

just politics. The Baptist, Salome,
that old king slicing his tender morsel.
The hen's love for the fox, the painter's the brush
switching its cutting arc
across the eye. The poor philosopher's
for truth, which keeps
changing its mind. This path's the only height

we can climb, sweating
not for the long view—for privacies,
gates set deep in the walls, early roses,
streets named for people
who are history. A pair of swans
scales the river below, and, to stop
the rising gaze, green hills, ruin at eye-level.

ROCCA MAGGIORE

The torture museum

Neck pulled back. Wrists tied. Weight pops
shoulder from socket. All disjunction.
So many Judgment Days. Hell absorbs us

as Heaven never can, though Heaven is all
martyrs dream of. Gold spraying the ceilings.
Choirs singing from memory in rows

measured out by goodness. If they succeeded
they are now escaped, not even scoured
blind by that light

> through which Hell resembles

this world, where even night curves into
a halo for the hills; in which a head knows
it's severed as it falls, if it's quick can think

I've lost my head—though it's the body lost, until
head comes to ground, pressing
earth that resonates for it, hooves

drumming nearer through the morning. The swift
rides currents under these ceilings; dove calls
ripple interior air. The head at least

got a clean break. These others surrender
to rack and pinion, stretched, dismembered, hung
upside-down, splayed, blades

sighing between their legs. Gravity. Blood

pumping air to heads wide awake
until the saw licks at navel or breast,
                                    so ravenous

it eats the saint alive.
                    Eyes back then
would have fixed upon him, ready to help
burn his hunger out.
                              If angels wore bodies

as armor, they'd be like us, turning
against temptation, folding
wings of moonlight, peacock feathers, ore

so airy no extra sinew holds them. Having
lucked into another
easy way. Kids, slick in leather,

smelling of Marlboros, lean over
that collection of switches, concentrate
desire, to open so far

air ignites their flesh, scrubs their bones
clean as these, dug up and boxed. *Desire
is so useless*, one says. The empty

grave glitters with tossed lire. From the tower
a countryside unrolls, tempts our eyes
beyond the courtyard's guillotine

weathered to rust and splinter, blade suspended
all these years, bed poised to lever the body
into its groove, the head carried away.

ADVENT

The National Gallery, early Dutch paintings

1. Before the merely human drew the eye,
these fields laid down rows ordered,

divinely touched. No perspective, what
you and I hope to gain: having made

five years together, we look toward fifty.
I have no way of knowing the world

didn't appear just so, frame-by-frame
flattened before heaven's deep idea

curving into infinity, beyond the canvas
over which the farmer shoulders his sack

away, sowing across the centuries.
Time deepens behind this day

whose surface we walk, fixed as pigment to fiber.
And how must the child have felt? Front and center,

enlarged, already out of his crib and holding
sway, he's perched on Mary's knee, his hand

perpetually lifted to bless, reaching into
a graver air, to pluck the ripened future.

2. I'd rather think of him laughing, or adrift
in utero, dreaming his infant's dream.

But first, annunciation, that euphemistic
consent—coercion, however divine,

must have been without sweetness, the instant
the head falls back, the neck arches,

and each consequent moment presents itself.
I wouldn't trust an angel, even restored

to luminosity, who would descend
to this, would resort to wearing wings

straight out of the sixties, shaded rainbows
distracting the eye from elaborated robes,

from the sly smile, the shaft of light piercing
Mary through the head. A dove

rides the beam, a superfluous gesture.
A floor, constellated with signs

of heaven she should have read, predicts this.
She might as well have raised her eyes to the stars

for all they'd tell her. If we believe the guard,
the restorer peeled the paint from the canvas,

fixed it to this one, cleaned off varnish, revisions
later artists imposed between our eyes

and the first breath. The guard likes us,
likes the way we lean, holding hands,

into each other, while Mary bows between
solid ground and the abstracted eye.

3. We search through all chaos for the saint.
Facing his own temptations: Old

Testament demons crawl toward the sky,
row cunning boats through clouds aswim

with creatures not of air, or sea, or earth,
but elemental to our human hearts—

fire roils through chambers, beats out
into extremities, though we hardly notice

most nights, at last together,
tired, looking for comfort. Until electricity

rises to the skin, your private weather
jolting against my fingers. How the taste

your lips bring to my lips tells me what happens
next is mystery. We are so

invisible to ourselves. We have learned
to trust what's in the air: frog-legged,

many-winged, long-fingered,
disarmingly lopsided. Look at this one,

levering from the ground. There's no escape;
his head's where his sex ought to be,

eyes bobbling upward. Anthony's
here at last, reading instruction manuals

beside a fallen tree. He's becalmed
to distraction, too fixed on God to see

the house lit within, the shifty sky,
woods ruffled with green. In all the scene

only he won't move—nor that city,
towered and walled against the miracle.

The Adoration of the Magi, The Annunciation, The Temptation of Saint Anthony